LUFTWAFFE AT WAR

Nightfighters Over the Reich

Early in 1945 the first Ju 88G-6s were captured by Allied forces in northern Germany. These USAAF officer are examining the AI radar installation of this G-6. Note the open entrance hatch and the two MG 151/20 weapons fitted in the weapon bulge together with two further MK 108 guns.

LUFTWAFFE AT WAR

Nightfighters Over the Reich

Manfred Griehl

Greenhill Books
LONDON

Stackpole Books
PENNSYLVANIA

Greenhill Books

Nightfighters over the Reich
first published 1997
by Greenhill Books, Lionel Leventhal Limited,
Park House, 1 Russell Gardens,
London NW11 9NN
and
Stackpole Books, 5067 Ritter Road, Mechanicsburg,
PA 17055, USA

British Library Cataloguing in Publication Data

Griehl, Manfred
Night fighters over the Reich. - (Luftwaffe at war;
v. 2) 1. Germany. Luftwaffe - History - 20th century
2. World War, 1939–45 - Aerial operations, German
I. Title
940.5'44'943
ISBN 1-85367-271-8

Library of Congress Cataloging-in-Publication Data

Griehl, Manfred
Nightfighters over the Reich/Manfred Griehl.
72p. 26cm. - (Luftwaffe at war; v. 2)
ISBN 1-85367-271-8
1. Nightfighter planes - History.
2. Airplanes, Military - Germany - History.
3. Germany. Luftwaffe. 4. World War, 1939–1945 –
Aerial operations, German. I. Title. II. Series.
UG1242.F5G7477 1997   96-39871
940.54'4943 – DC21   CIP

Designed by DAG Publications Ltd
Designed by David Gibbons.
Layout by Anthony A. Evans.
Edited by Ian Heath.
Printed and bound in Singapore.

The photographs in this book were taken under
wartime conditions. Some are not of the highest
quality, and are included because of their
extreme rarity.

NIGHTFIGHTERS OVER THE REICH

At the outbreak of the Second World War on 1 September 1939 only five small German *Staffeln* (Squadrons) had been designated as dual day and nightfighter units. These were 11 (*Nacht*) *Staffel* of *Lehrgeschwader* (Training Wing) 2; 10 (*Nacht*) *Staffel* of *Jagdgeschwader* (Fighter Wing) 2; 10 (*Nacht*) *Staffel* of *Jagdgeschwader* 26; and 10 and 11 (*Nacht*) *Staffeln* of *Jagdgeschwader* 72. These units were based respectively at Düsseldorf, Fürstenwalde near Berlin, Garz on Usedom Island, Mannheim-Sandhofen, and Stuttgart-Echterdingen. Because JG (*Jagdgeschwader*) 72 was never completed, later in 1939 its *Staffeln* became part of JG 52. These *Nachtjagdstaffeln* (nightfighter squadrons) were used to gain night fighting experience and to train as many pilots as possible in this new operational role.

The RAF's first night-time raids over Germany caused little damage, not least because at this early stage of the war bombers lacked the technology to find their targets in the darkness. However, the fact that the raids had occurred at all, and the realisation that the scale of such operations was likely to grow, resulted in the *Luftwaffe* providing Bf 109 'nightfighters' to IV/JG 2 (IV *Gruppe* of *Jagdgeschwader* 2), the component fighter units of which were to concern themselves purely with night flying.

The first, but still unofficial *Nachtjagdgruppe* (nightfighter Group) became *Oberleutnant* Wolfgang Falk's I/ZG 1 (I *Gruppe* of *Zerstörergeschwader*, or Destroyer Wing, 1). Based at Aalborg near the Danish border, *Gruppenkommandeur* Falk's unit, equipped with Bf 110C *Zerstörer* (destroyer) aircraft, had taken part in the campaigns in Denmark and Norway in April 1940. Aalborg was one of the early Freya radar installations, designed to provide early warning of RAF raids towards Northern Germany. Despite every effort, however, even with the aid of radar-assisted ground control Falk and his men were unable to intercept the enemy raiders, and claimed no night-time victories.

There were no further developments until June 1940, when parts of both I/ZG 1 and IV/ZG 26 were combined to form the *Nacht- und Versuchsstaffel* (nightfighter and Experimental Unit), to commence new operations from Düsseldorf. Then on 17 July 1940, *Oberst* Josef Kammhuber was ordered to organise a modern, well-equipped nightfighter force to protect Germany's industrial centres from steadily increasing night-time bombing raids. At that time the future *Generalleutnant* had no idea how to apply radar to the interception of enemy aircraft. Furthermore the *Nacht- und Versuchsstaffel*, together with a few *Staffeln* of IV/JG 2, still flew only Bf 109 fighters. Kammhuber nevertheless set to work, and just three days later *Nacht- und Versuchsstaffel* became part of I/NJG 1 (*Nachtjagdgeschwader*, or nightfighter Wing, 1), which was formed mainly from I/ZG 1 and would later receive heavy Bf 110 fighter aircraft. The Wing's II *Gruppe* consisted initially of three *Staffeln* and soon received a fourth, the former 4 (*Zerstörer*)/KG 30 (KG standing for *Kampfgeschwader*, or Bomber Wing). Later, in summer 1942, II/NJG 1 became the nucleus of NJG 2, and in winter 1941–2 the *Ergänzungsstaffel* (Supplementary Squadron) of NJG 1 and the *Ergänzungsgruppe* of ZG 26 were combined as IV (*Ergänzung*)/NJG 1, which was later redesignated as III *Gruppe* of *Nachtjagdschule* (nightfighter School) 1. After this unit had left NJG 1, II/NJG 2 formed a new fourth *Gruppe*. II/NJG 1 was classified as the *Luftwaffe*'s first *Fernnachtjagdgruppe* (Long-Range nightfighter Group), with orders to undertake intruder missions against RAF bomber bases all over Great Britain.

Until the end of July 1940 II *Gruppe* operated two squadrons equipped with early Ju 88C-1 nightfighters, modified from Ju 88A-1 bombers, and a limited number of Do 17s, consisting of Z-7 'Kauz I' and Z-10 'Kauz II' models. Of the Ju 88C-1s and Do 17Z-7s very few had been modified in any way. Ju 88C-2s, of which sixty-five were modified A-5 series aircraft, and nine rebuilt Do 17Z-10s comprised the new group's first operational equipment.

NJG 2 was formed in autumn 1940 and was intended to become a long-range nightfighter formation. Its first *Gruppe* was the former II/NJG 11, while its second was built up during winter 1941–2 and was equipped with Ju 88C nightfighters. From summer 1942, II/NJG 1 was redesignated IV/NJG 3, NJG 3 having been established in October 1940, when V (*Zerstörer*)/LG 2 became I/NJG 3.

On 17 October 1940 Josef Kammhuber was promoted to *Generalmajor*. He had introduced a new kind of air-defence system along Germany's western frontier, where the vitally important Lower Rhine and Ruhr industrial zones were situated. Kammhuber divided the sky into air defence regions referred to as 'boxes', each having a length of 40 km and a width of some 25 km. Only nightfighters were permitted to fly within these Along the forward perimeter of each of these boxes sound detectors were installed to provide advance warning of approaching enemy aircraft. Behind these were located huge searchlights. Each box also had a radio beacon, which vectored its allocation of nightfighters against individual attacking bombers. If the nightfighter lost contact with its target, it returned to the fixed beacon, either to make another attempt or to wait until it was allocated another target. More and more radar-equipped units were assigned to this system, these being installed along the European coastlines and throughout Germany's industrial and urban heartland, especially around Hamburg, Berlin and the Ruhr and in the vicinity of Frankfurt am Main. (At a later date, when each box comprised three radars and a ground control system, it became known as a *Himmelbett*, or 'four-poster bed'.) In addition, in northwest Germany the tactic known as *Helle Nachtjagd* was adopted, in which aerial flares were dropped to illuminate approaching RAF bombers, enabling the nightfighters to pick them out and destroy them.

Because searchlights needed clear weather, and an element of luck, to pick out their targets by using sound locators, while the nightfighters had to find their targets flying blind, assisted only by radio aids, initial results were not particularly encouraging. It took Germany many months to introduce suitable technological improvements, but in autumn 1940 the Würzburg radar became available, capable of giving precise data regarding the range, speed and height of enemy aircraft, though nowhere nearly enough of these radar systems were ever produced. This and the larger Freya radar were used to gain sufficient advance warning of the enemy's approach for the searchlights and anti-aircraft guns to be ready to locate hostile aircraft as they penetrated the air defence box. Tracking and pursuing aircraft by means of radar became known as *Dunkel Nachtjagd*. Like the guns, nightfighter pilots had their activities co-ordinated by the men under the command of *Generalleutnant* Dr-Ing Wolfgang Martini, who was responsible for radar and radio equipment and was promoted to *General der Luftnachrichtentruppe* on 20 September 1941.

Despite the tactical potential of I/NJG 2's long-range nightfighter operations, Hitler called a halt to further activities over England after some twenty RAF bombers had been shot down. The unit was subsequently transferred to the Mediterranean theatre, where its pilots racked up more than 140 air victories. At this stage it was suggested that the *Himmelbett* box concept be improved by using radar and larger searchlights units. The subsequent invention of the *Seeburg-Tisch* enabled nightfighter activities to be co-ordinated more effectively with those of AA batteries and radar installations in their sector. By the end of December 1941 a line of six boxes had been established between the mouth of the Rhine and the Danish border. Additionally *Helle Nachtjagd* zones were built up between St Trond and Stade in order to provide a second line of air defence. In combination with nightfighter zones round Köln-Düsseldorf, Bremen, Hamburg or Berlin, both *Helle* and *Dunkel Nachtjagd* were believed to be effective in either stopping flights of incoming RAF bombers completely or in hindering most of them from releasing their bombs accurately.

During 1940 the RAF lost 342 of its aircraft over Germany; in the following year some 1040 were either shot down by nightfighters or AA guns, or else crashed in consequence of mechanical failure, with many of these losses occurring in engagements with German nightfighters. The improvement resulted from the *Luftwaffe* now having sufficient aircraft available over the Reich to respond to every enemy raid. In May 1942 Kammhuber's XII *Fliegerkorps* (Air Corps), which had been responsible for nightfighter defence since 9 August 1941, had charge of three *Jagddivisionen* (Fighter Divisions), commanded respectively by Major Generals Kurt-Bertram von Döring and Walter Schwabedissen, and *Oberst* Werner Junck. NJG 1, together with two signal regiments (numbers 201 and 211), belonged to 1 *Jagddivision* (HQ Deelen), 2 *Jagddivision* (HQ Stade) was responsible for the operations of NJG 3 and three signal regiments (numbers 202, 212 and 222), while the third (HQ Metz) commanded II/NJG 2, III/NJG 4, and signal regiments numbers 203 and 213.

With the introduction of more powerful RAF bombers, which appeared over the Reich in ever larger numbers, ordinary Bf 110 and Ju 88C nightfighters were found inadequate to their task. Improved versions of both aircraft were therefore put in hand. In the resulting Bf 110G-4, with additional fixed oblique weapons in the rear cabin (Schräge Musik), and Ju 88G-1, propelled by two BMW 801 engines, Germany found two reliable aircraft to defend the Reich. Simultaneously with their development during summer 1943, production of He 219A and Ju 188S/R nightfighters was also considered. Because of its higher performance values, a few prototypes of the He 219, along with six pre-production aircraft, were sent to the *Fronterprobungsstelle* (evaluation unit) at Venlo, and during *Oberstleutnant* Streib's first mission using an He 219 he shot down five bombers.

Despite its established operational success, General Kammhuber's air defence system became obsolete on the night of 24/25 July 1943, when RAF heavy bombers used 'Window' for the first time, consisting of clouds of aluminium foil which confused radar reception and made it impossible to take precise readings. This simple expedient rendered it very difficult to make effective use of the data provided by German radar — both ground-based Würzburg installations and the airborne FuG 202 and FuG 212 'Lichtenstein' systems installed in most twin-engine nightfighters. Although Kammhuber tried to increase the number of nightfighters and *Himmelbett* sections to compensate, and new 'Window'-proof hardware was eventually introduced, it took months for the *Luftwaffe* to regain some of its lost initiative. Kammhuber was meanwhile promoted to Commanding General of all nightfighter units on 15 September 1943, and was posted to Norway on 20 November 1943, where he was appointed to *Luftflotte* 5.

His former *Fliegerkorps* was now provided with two additional *Jagddivisionen*, the fourth being commanded by *Generalmajor* Joachim-Friedrich Huth, with its HQ at Döberitz near Berlin, and the fifth by *Oberst* von Bülow, situated at Schleissheim near Munich. Under 1 *Jagddivision* were placed *Gruppen* I to IV of NJG 1; 2 *Jagddivison* was responsible for I to IV/NJG 3; and just one *Gruppe*, II/NJG 4, was situated in the command area of 3 *Jagddivision*. Because of its importance six nightfighter groups were stationed around Berlin (I and II/NJG 2, I and III/NJG 4, and I and III/NJG 5). The Munich area was protected by smaller forces belonging to I/NJG 6 and the NJG 101 training formation, where the so-called *Einsatzschwärme* (operational swarm) had been taught.

Following the partial failure of the old *Himmelbett* system new tactics were evolved under the codenames *Zähme Sau* (Tame Boar) and *Wilde Sau* (Wild Boar). In *Zähme Sau* attacks radar was used to guide the nightfighters to those areas where 'Window' was at its thickest, after which they would dog the bomber stream for as long as possible. In the *Wilde Sau* attack, evolved by *Oberstleutnant* Hajo Herrmann, nightfighters instead picked their targets out by looking for their silhouettes, either against the the glow of the burning towns beneath them, or against a thin layer of low cloud, which reflected the light of the fires and was sometimes further illuminated by means of searchlights. In addition *Beleuchter* (illuminator) units facilitated nightfighters attacks by dropping flares to illuminate the upper regions of the sky. Anti-aircraft units located within the attack zone were ordered to fire only to a certain altitude, in order not to endanger *Wilde Sau* fighters. These tactics were employed for the first time by an experimental nightfighter *Kommando* (detachment) on 3/4 July 1943, when RAF bombers attacked the city of Cologne, and resulted in twelve bombers being shot down. This success led to the formation of several larger units, *Oberstleutnant* Herrmann being ordered to establish 30 *Jagddivision*, which had under its command three *Wilde Sau-Geschwader*, JGs 300, 301 and 302. This unit was disbanded on 16 March 1944, the remaining *Staffeln* being used as bad-weather fighters over the Reich until April.

During this period nearly forty He 219s were delivered to I/NJG 1, stationed at Venlo, which unit brought down about 100 British aircraft. Some more He 219s were operated by 2/NJGr (*Nachtjagdgruppe*) 10, based

at Finow. Other than these improved aircraft, most of the *Luftwaffe*'s nightfighter inventory consisted of Ju 88C-6s and G-1s. However, by summer 1944 more and more of the faster G-6s were being delivered to replace older versions of the famous Ju 88 nightfighter. By that time only a few Bf 110G units still existed. Of two advanced piston-engine nightfighters, the Ta 154 did not reach operational status until May 1945, while the Ju 388 probably never did.

In November 1944 all nightfighter units were subordinated under the command of 1 *Jagdkorps*, which had been led by *Generalleutnant* Josef Schmid since 15 September 1943. This was part of *Luftflotte Reich*, commanded by *Generaloberst* Hans-Jürgen Stumpff. To 1 *Jagddivision*'s NJGr 10, an experimental unit equipped with He 219s and late Ju 88 conversions, belonged *Fliegerführer Ostpreussen* (Fighter Leader East Prussia) and a second fighter leader, responsible for Schlesien. I and IV/NJG 5, together with crews from II/NJG 5 and instructors from NJG 102, operated under this command. The second fighter division at Stade and the subordinated Fighter Leader Denmark, *Oberst* Vieck, operated NJG 3, IV/NJG 2, 1 *Staffel* NJG 5, and 1 *Staffel* NJG 11. In 3 *Jagddivision* were concentrated parts of NJG 1, 2 and 4, together with smaller formations of NJGr 10 and NJG 11. To 7 *Jagddivision* belonged major parts of NJG 6 and I/NJG 101. Finally, 8 *Jagddivision*, based at Wien Koblenzl, was responsible for the defence of former Austria. There were only three *Gruppen* based in eastern Germany (III/NJG 6, II/NJG 100 and II/NJG 101), these operating alongside 4/NJG 100 and an Hungarian unit (NJSt 5/1).

By the end of November 1944 most of the front line units (III to IV/NJG 1, NJG 2 to NJG 6, and NJG 100) flew heavy Ju 88G-1 and G-6 nightfighters. Most of the fast He 219As were concentrated on the airfields of I and II/NJG 1, but a few belonged to NJGr 10, which also operated Bf 110s, Ju 88Gs and a few single-engine nightfighters. NJG 11 and all three of the *Wilde Sau* units (JG 300 to 302) were equipped with Bf 109G and FW 190 fighters. A mixture of Bf 110s and Ju 88Cs and Gs saw operational service with I/NJG 101, while III/NJG 101 and all three groups of NJG 102 had Do 217N and Ju 88 heavy fighters on their inventories. Besides these aircraft some experimental use was made of a few Me 410 aircraft, serving with I/KG 51. By the end of the war many of the Ju 88G-6s were used instead for low level attacks against Allied columns advancing on Hamburg, Munich and Berlin, while aerial activity of any kind was severely limited by a lack of aviation fuel: in the absence of both fuel and tractors the Ju 88Gs had to be towed to the end of the runway by oxen, or even by hand.

During summer 1944 Arado had developed a new jet-propelled two-seat nightfighter, the Ar 234 'Nachtigall' (Nightingale), of which the first production aircraft were anticipated to be available by November. However, the initial full-size mock-up took more than four months to build, and it was early November 1944 before the first Ar 234 prototype (c/n 140145) was delivered for modification. Simultaneously with this plans for the construction of many more Ar 234 (P-series) nightfighters, along with some modified variants (the Ar 234C-3/N and C-5/N) were put in hand. On 23 February 1945 the first experimental Arado crashed as a result of pilot error, killing both *Hauptmann* Bisping and *Hauptmann* Vogel, the radar operator. A month later a new evaluation unit, called *Kommando Bonow Ar 234*, was set up. Under the command of *Hauptmann* Bonow, former leader of *Kommando* 388, three Ar 234s underwent nightfighter evaluation but without any operational success. This small unit was destroyed by an Allied raid on its airfield near Oranienburg, where a few Me 262 jet-propelled nightfighters had also been stationed, together with a Do 335 with FuG 220 antennae.

Besides a few Ta 154 piston-engined nightfighters which were tested early in 1945, a few Me 262 jets were also modified for evaluation trials with radar equipment. The chief of technical armament (*Chef Technische Luftrüstung*) had ordered the construction of some of these new fighters by the end of January 1945, but technical difficulties delayed their entry into operational use for several weeks. Despite the fact that development of both the Ar 234 and the Me 262 showed only slow progress, in February 1945 a heavy nightfighter with a modern FuG 240 antenna, and propelled by two or more jet engines, was ordered by the RLM (*Reichsluftfahrtministerium*, the German Aviation Ministry). Because the highly reflective glazed canopy of the Ar 234 was deemed dangerous, alternative designs were sought from other aviation companies, including Messerschmitt, Horten and Junkers, whose engineers, seemingly oblivious to the fact that Allied armies were already marching on Berlin, set about their task.

On 20 March 1945 a *Führer-Befehl* (a personal command from Adolf Hitler himself) ordered the production of all single-engine piston fighters to cease, with the single exception of the fast Ta 152, which was needed to protect German jets when they were at their most vulnerable, during landing. Only a few days later, on 26 March 1945, Hitler ordered the construction of a new type of night and bad-weather fighter, equipped with two jet turbines under the wings and a piston engine fixed in the rear fuselage to increase the aircraft's theoretical operational range. Some two-seat Do 435 'Pfeil' (Arrow) rear-engine nightfighters were also to be produced for evaluation purposes, but only a single experimental prototype ever appeared, and this was captured by the Red Army at Oranienburg.

More than 100 all-weather jet or rocket-propelled designs were finished by the end of the war, but only the jet-propelled Me 262 nightfighter was ready to enter limited operational service early in 1945. Its development had begun in September 1944, and on 11 December the *Oberkommando Luftwaffe* permitted the establishment of two experimental units, which would be equipped with a mixture of Ar 234 and Me 262 nightfighters. After first trials early in 1945 using an experimental Me 262A-1a (c/n 170056), the first operational mission was flown on 15 February. Most of the aircraft of *Kommando* Welter (subsequently 10/NJG 11) were Me 262A-1a jet fighters; later it was proposed to build a two-seat Me 262B-1a/U1 nightfighter, carrying FuG 220 radar. By 21 March 1945 this small unit had scored some nineteen air victories, with another seven by the end of the month. By the end of the war more than forty Mosquitos and four-engine bombers had been shot down by 10/NJG 11, which eventually lost most of its own aircraft through technical problems, only a few being damaged or shot down by the Allies. On 7 May 1945 this *Staffel* surrendered at Schleswig-Jagel.

The first Bf 110 nightfighters were coloured black overall and featured the 'England Blitz' badge of NJG/1 on both sides of the forward fuselage. Later most Bf 110 nightfighters had a mottled grey camouflage on the fuselage, fin section and engine nacelles. The upper surfaces usually showed a splinter pattern of dark and medium grey.

Above: After flying its first missions in the splinter camouflage common to all day destroyers, most Bf 110Cs of NJG/1 were painted black. Many of the first nightfighter crews were recruited from experienced personnel belonging to long-range reconnaissance, bomber and transport units, plus flight instructors and former Lufthansa pilots.

Below: During winter 1940/41 more than 60% of all Bf 110s built went to nightfighter formations. However, output figures were low and few additional units could be raised. The number of well-trained crews was also insufficient: on 4 January 1941 only sixteen crews were available with the appropriate training to fly nightfighter missions.

The Ju 88G-1 was one of the most successful German nightfighter designs to be evaluated at the *Erprobungsstelle der Luftwaffe* (*Luftwaffe* evaluation unit) at Werneuchen in central Germany, where all kinds of radar installations were installed and tested. The aircraft shown was one of the experimental Ju 88s equipped with FuG 220 radar.

This BMW 801-powered Ju 88G-1 belonged to IV/NJG 3, a unit where new pilots were readied for combat missions. Note the horse-drawn wagon on the runway beyond; owing to a lack of fuel *Luftwaffe* units used horses to transport ammunition and spare parts around their airfields.

This single-seat Bf 109 nightfighter was captured near Cologne by American ground forces early in 1945. Some nightfighters such as this were sent up to intercept the RAF's fast Mosquito reconnaissance aircraft, but most such missions were failures. Among the few pilots who could claim such kills to their credit were *Hauptmann* Karl-Heinrich 'Nasen' Müller (I/NJG 10) and *Oberleutnant* Kurt Welter (10/JG 300). However, these all took place in daylight.

To evaluate the possibilities of a single-seat nightfighter (*Einsitziger Nachtjäger*) this Bf 109G-6/N was tested at Werneuchen. Along with a few other aircraft the Bf 109 was used to discover the range of FuG 217 and FuG 218 radar installations.

Only one single-seat Me 262A-1a, with the production number 170056, was fitted with an AI-radar installation. Despite all the stories that have been told, this aircraft never saw operational service with a *Nachtjagdgeschwader*, though it was flown by *Leutnant* Welter for some time to test its night fighting potential when equipped with FuG 218 radar. However, Welter's 10/NJG 11 did later fly Me 262A-1a nightfighters which differed very little from the day fighter variant.

Above: All the lessons learned with the Me 262 V056 were subsequently applied to Me 262B development. The sole Me 262B-2, possibly assembled early in 1945, never got as far as flight evaluation before the war ended. The damaged V056 was captured by American ground forces at Lechfeld, where many other experimental prototypes also fell into Allied hands at the end of April 1945.

Below: Many antennae were tested at Lechfeld. Between December 1944 and April 1945 there were several meetings between high-ranking officers to discuss three important two-seat nightfighter designs, those of the piston-engined Do 335, the jet-powered Ar 234 and Me 262; but no decision was reached.

Above: An early German night fighter, with ace Ludwig Becker in the cockpit. The Do 17Z-7, called 'Kauz I' (Screech Owl I), was the first twin-engine Dornier-built heavy fighter aircraft. Most of them were operated by 4 *Staffel* of *Nachtjagdgeschwader* (NJG) 1. The Do 17Z-1 was propelled by two Bramo 323 P piston engines, each rating some 1000 HP, and was capable of a maximum speed of some 410 km/h. Becker scored 46 night victories.

Right: The first base used by the still-experimental Do 17Z-7 delivered to II/NJG 1 was the *Luftwaffe* airfield near Düsseldorf, which is where this picture was taken. Due to the lack of supporting infrastructure, its first interception mission over the western part of the Reich did not end in success. The armament of Do 17 nightfighters was later improved by the installation of additional fixed armament in the forward fuselage.

Above: The next step in the evolution of an improved version of the Z-7 was the Do 17Z-10, which could reach a maximum speed of 428 km/h at an altitude of 5900 m. Most of these aircraft only saw limited service with front line units before being handed over to NJG 101, a training unit. This Z-10 is equipped with an active infra-red device, called 'Spanner-Anlage'. This was tested for many months but was never introduced in large quantities because of its limited range.

Below: This Do 215B-5 belonging to 5 *Staffel* of NJG 2 was photographed in May 1942. It carries an early version of FuG 202 radar, which had a range of only some 1800 m. A lack of spare parts and losses in action over the Reich meant that Do 215B-5 numbers remained low, and the type was eventually withdrawn from active service as more Bf 110 and Ju 88 nightfighters became available.

Above and below: This Do 17Z-10 (R4+HK) belonging to II/NJG 2 crashed during late summer 1941 in the northern Netherlands as a result of enemy action. It is fitted with four fixed MG 17s, which were installed along with one or two forward-firing MG 151/20 guns in the nose. This heavy armament made an easy job of destroying all kinds of RAF heavy bombers at close range.

Above: The Do 215B-5 was capable of some 470 km/h and was later improved by the use of a FuG 202 'Lichtenstein' radar system, which had a range of 3500 m. The first Do 215B-5 to score a kill with the help of 'Lichtenstein' radar was commanded by *Oberleutnant* Lent, whose aircraft was provided with two additional MG FF beneath the cockpit section, increasing its fire-power to four 20 mm weapons and four smaller MG 17s, all fixed in the nose. The aircraft depicted belonged to II/NJG 2.

Below: The next phase in Do 217 evolution was the development of a night fighter based on the medium Do 217E-2 bomber, to fulfil the role of a close-area heavy night fighter. The first Do 217J-1 (RH+EH, s/n 1134) was tested at Tarnewitz proving ground. However, all the early Do 217 nightfighters to be built were said to be too slow to make successful fighters. Their experimental use as heavy day aircraft destroyers similarly failed through poor performance.

Above: During the final phase of the Do 217J's operational career the fixed armament of many was altered from MG FF to the more effective MG 151/20 cannon seen here. The war ended before the planned installation of air-to-air missiles could take place, although tests were carried out with one Do 217 night fighter carrying Hs 298 missiles for use against heavily-armed targets.

Below: After the improved Do 217N night fighter became available, most J-1s and many J-2s were handed over to NJG 101 and other training formations. In addition to their training role, the flying instructors of NJG 101 saw action with the so-called *Einsatzschwärme*, small units of up to six aircraft established to intercept enemy formations by night. However, these achieved few victories.

Above: This is the fourth prototype of the Do 217N-1, the N-04 (s/n 1404, code GG+YI), powered by two DB 603 A-1 engines. The installation of these more powerful water-cooled piston engines was the most important difference between the J- and N-series, and made possible the fitting of the dorsal *Schräge Musik* ('oblique music') installation, consisting of two or four 20 mm cannon, most commonly two. These improvements brought the early N-series aircraft to Do 217N-2/R22 standard.

Top right: This Do 217N-07 (GG+YG) is still operating with FuG 202 AI-radar. The aircraft was later improved by a better wireless system and a FuG 25a friend-or-foe installation. Some early aircraft were also tested with FuG 212 radar. The use of FuG 220 in Do 217s was abandoned because the system was needed to equip Bf 110 and Ju 88 nightfighters for the defence of the Reich.

Right: This Do 217N-2 saw operational use between February 1943 and September 1944, when it was used for training purposes. The defensive turret in the rear part of the cockpit was removed and an aerodynamic fairing was installed over the ventral gun position step. Despite these alterations, this refined version of the Do 217 remained too weak to enjoy an active operational career.

Below: A four-seat Do 217J fighter. These were normally equipped with four small antennae belonging to a FuG 202 or FuG 212 radar installation. In September 1943 a few Do 217Js were fitted with FuG 220 (SN-2) radar, which had larger antennae, but only a few saw action. Their armament did not differ from the series J-2, but unlike the J-1 the rear bomb bay was sealed off.

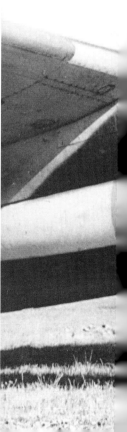

Top left: This Do 217N-2 of III/NJG 1, shot down by a B-24's rear gunner, belonged to *Feldwebel* Kustusch, an experienced flying instructor who flew in action as well as training new pilots. Some of his missions were carried out with pupil crews over Hungary in summer 1944. He survived the war and now lives in a little town near Frankfurt.

Above: The cockpit interior of a Do 217N-2 belonging to 9 *Staffel* of NJG 6. The aircraft and crews of this squadron had been transferred to NJG 6 after enduring long and costly operational careers all over the Reich during which many aircraft had been lost to Allied fighters.

Left: The Do 217N-2, a rebuilt N-1/U1. It saw action from May 1943 and could reach a maximum speed of some 510 km/h at an altitude of 6000 m. Operational range was 2100 km. Although its fire-power was greater than that of a Bf 110 night fighter, manoeuvrability was lower than that of other Dornier aircraft or Ju 88C and G nightfighters.

Top left: With an offensive armament comprising two MG FF and two MG 17, the Bf 110F-2 could attain a maximum speed of 540 km/h at 7000 m. The endurance of first series Bf 110 nightfighters was rather poor, but following further development the G series was to become a formidable weapon in the hands of an experienced crew. The aircraft shown, equipped with two 300-litre drop tanks, is that of *Oberleutnant* Hans-Karl Kamp, *Staffelkapitän* of 7 *Staffel*, NJG 4. Kamp scored twenty-three victories before he was killed on 31 December 1944.

Left: II/NJG 3 operated several Bf 110C-1, C-2 and C-4 aircraft, that depicted having seen action all over Europe including intruder sorties over England. The Bf 110C was powered by two DB 601 A-1 engines and was produced at Haunstetten near Augsburg, Bavaria. The development of the Bf 110 was dispersed all over Germany to avoid the total destruction of production lines by Allied air raids. Note the row of flags painted beneath the cockpit, signifying countries in which the pilot had fought.

Above: While returning from a sortie this Bf 110 was forced to belly land not far from Twenthe airfield after receiving several hits from the rear gunner of a Halifax or Lancaster over the Cologne—Düsseldorf region in 1943. Both engines broke out off their fittings, and the main structure of the fuselage was damaged. Both crew members survived.

Right: During early battles over the Reich it was found that even Whitley, Blenheim and Wellington bombers did not constitute easy prey. However, the four-engine RAF heavy bombers which succeeded them — the Stirling, Halifax and Lancaster — could be truly formidable opponents. The first Short Stirling was shot down on 10 April 1941 by the crew of *Feldwebel* Scherfling (7/NJG 1), while the first Halifax was brought down by *Oberleutnant* Eckardt of NJG 1.

Above: This crashed Bf 110 belonged to the staff of II/NJG 1. Prior to the delivery of the more powerful Bf 110G this unit was equipped with C-2, C-4, and D-0 to D-3 aircraft, plus some F-4s. Possibly this aircraft was shot down by one of the long-range RAF fighter crews which operated over north-west Germany.

Below: Another downed Bf 110 night fighter, this time of 7 *Staffel* NJG 1, lost after one of the engines was set on fire by an RAF air gunner. The pilot carried out an emergency landing without undercarriage. Many NJG 1 missions were carried out from Deelen near Arnheim, Laon, Bönnighardt — where the *Geschwaderstab* (wing staff) was established — and bases near Düsseldorf, Krefeld and Münster.

Right: A Bf 110 of NJG 2 (code R4+). Formed in September 1940, and eventually claiming a total of more than 800 victories in night and day actions, NJG 2 was commanded for some time by Prince Heinrich Sayn-Wittgenstein and Günther 'Fips' Radusch. I/NJG 2 was formed from II/NJG 1 and 1/ZG 1, while II *Gruppe* was made up from 4/NJG 2 and parts of ZG 2. During March 1942 a third *Gruppe* was established, but on 1 October 1942 this was redesignated II/NJG 2. The *Geschwader*'s last major missions were flown in 1945, from Twenthe and Vechta.

Bottom right: During winter, heating systems were vital to the preparation of nightfighters for their next sortie. Hot air was used to warm the piston engines and to de-ice other important parts.

Above: This Bf 110G-4 carries a modified day fighter camouflage and still has the small fins of a Bf 110F aircraft. It belonged to 7 *Staffel* of NJG 5 (2Z+). Being engaged in long-range defensive missions, this aircraft is provided with two 300-litre drop tanks. After commissioning, 7/NJG 5 saw operational service until the end of the war under the command of *Hauptmann* Leuchs and *Hauptmann* Helmuth Schulte.

Below: D5+LT and another Bf 110G-4/R1 of III/NJG 3 above the clouds over central Germany, heading back to their base. They are equipped with FuG 202 'Lichtenstein BC' AI radar. The G series could maintain a speed of 525 km/h over considerable distances, though with additional armament performance dropped to 485 km/h.

Right: A Bf 110F-4 of 14 *Staffel* of NJG 5. Until April 1944 a few of these were used for night-time bombing. Normally they carried two MG 151/20s and four MG 17s in their nose sections. Additionally a *Schräge Musik* installation comprising two MK 108s could be installed in the rear section of the cockpit. A few Bf 110F-4s were rebuilt to receive two more cannon. In East Prussia these aircraft were flown with FuG 212 radar until April 1944.

Below: Bf 110G-4 VQ+KL (s/n 5538). En route from Gütersloh to its operational unit at Mainz-Finthen and crewed by *Leutnant* Lange and *Gefreiter* Achterberg, it crashed some 1.5 km north-west of Pruntrut. The inexperienced crew lost radio contact shortly after take-off, overshot Finthen by about 275 km, and landed with the wheels up, seriously damaging the aircraft.

Left: This Bf 110G-4 is partially sheltered from low level attack by its proximity to the edge of a wood. The three-seat G-4 was armed with two 30 mm machine cannon and two 20 mm cannon in the lower part of the forward fuselage. If service as a heavy fighter-bomber was called for, modification kit R2 allowed the fitting of a bomb rack below the fuselage, capable of carrying several different bomb loads.

Bottom left: This Bf 110G-4/R1 was used as an all-weather heavy destroyer. Beneath the fuselage a 37 mm *Bordkanone* BK 3.7 (Flak 18) was housed in a ventral fairing interchangeable with an under-fuselage ETC bomb rack. Instead of two MK 108s this aircraft was equipped with two MG 151/20 cannon. Two 300-litre drop tanks were also carried. The photo was taken at Munich-Riem in summer 1944. Note the words *Keine Bombe!* (not a bomb) on the underwing fuel tank.

Top right: In May 1945 a pilot and mechanic of NJG 1's staff stole this Bf 110G-4 night fighter (DV+IZ, s/n 5678) and escaped to neutral Sweden, where they landed in a meadow. The crew was interned and thereby avoided becoming prisoners of the Allies. The aircraft was subsequently evaluated by Swedish specialists in order to find out more about German AI-radar installations.

Right: The heavy MK 108 armament in the nose section of a Bf 110G-4. The aircraft could reach a maximum speed of some 485 km/h when fitted with this. These weapons were really too light for service against the more powerful of the various Allied bombers they faced over the Reich, but could destroy all manner of lighter combat aircraft with one to three direct hits.

Left: The 20 mm cannon ammunition storage in the nose of a Bf 110G-4. The explosive head of just one round could cause considerable damage, but on the whole more than three hits were required to guarantee the destruction of a four-engine bomber.

Bottom left: This Bf 110G-4 was operated by *Feldwebel* Kustusch from bases near Halle and Dresden in 1943. It was used for conversion training by NJG 2. Delivered only shortly before the end of the war, when handed over to the Allied forces it still had no unit codes. Its flame-dampers had been removed and the radar antennae rebuilt.

Right and below: During winter 1943–4 only a few Bf 110G-4s were equipped with conversion kit M1, which consisted of two more MG 151/20 guns plus ammunition storage container, both fixed under the fuselage. The radar antennae comprised a FuG 212 C-1 together with a FuG 220 (SN-2b). This equipment widened the angle within which targets could be located and, along with an SN-2 AI radar system, could scan a larger area.

Left: This picture of a Bf 110 G-4, fitted with flame-damped MG 151/20 weapons instead of the more common MK 108, was taken in summer 1944. The aircraft was flown by crews of 5/NJG 5. From September 1943 the provision of many of NJG 5's G-4s with SN-2 radar systems enabled them to locate Allied bombers at long range.

Below: Allied air superiority necessitated the use of as much concealment as possible. This Bf 110G-4 belonged to NJG 1 and was photographed near an airfield in northern Germany during summer 1944. Despite German attempts to conceal aircraft many nightfighters were found and destroyed by the low level ground attacks of long-range fighter-bombers.

Right: A close-up view of R4+BP, flown by 6 *Staffel* of NJG 2. It was used by *Feldwebel* Kustusch, who later became an *Oberfeldwebel* during the closing days of the war. Lack of fuel meant that many night fighter units remained grounded for most of their operational careers. From early in 1945 it was only possible to carry out a handful of sorties, flown by aces drawn from one or other of the much-weakened units.

Bottom right: Another close-up, of an aircraft which was subsequently lost over central Germany a few days later, on 4 October 1944. Having taken part in an air-defence sortie, it crashed after an engine failed when approaching Wilhelmshaven.

Above: R4+BP was flown by *Feldwebel* Kustusch, who claimed two air victories. Its defensive armament consisted of one MG 81 Z gun, operated by the wireless operator sitting in the rear cockpit. Many of the aircraft used by II and III/NJG 2 had a bright, nearly white camouflage called *Lichtgrau* 76 on both sides of the fuselage. The white crosses normally used on nightfighters inevitably disappeared against this background.

Below: At one of the former *Luftwaffe* airfields near Prague — possibly Prague-Gbell or Prague-Rusin — an abandoned, damaged Bf 110G-4 night fighter is examined by locals in summer 1945, following the end of the war. Behind it the wreckage of a Do 17 can just be seen.

Above: The wreckage of one of the special trailers that the *Luftwaffe* used to move damaged aircraft sections to the nearest railway station. The widespread, albeit damaged, railway network at first made it relatively easy to send wreckage back to a maintenance depot. Each of the main types of combat and transport aircraft was designated one or more factories for repair and recovery operations. The restoration of a damaged aircraft also provided the opportunity to upgrade it with more powerful weapons or better engines.

Below: This Bf 110G-4 (D5+CS), part of 8/NJG 3, is being prepared for action by its ground crew, nicknamed 'black men' on account of their black overalls. 8 *Staffel*, formed from 5/ZG 76 in November 1941, was initially equipped with Bf 110C-4, D-3, F-4 and G-4 nightfighters, but by the end of the war most of its aircraft were Ju 88G-6s. Besides these a few Ta 154s and He 219s were flown from bases in Northern Germany.

Above: This long-range night fighter, a Bf 110G-4, was captured by American ground forces in April 1945 during their advance into the heart of Germany. It belonged to 4/NJG 100, which became *Kommando*/NJG 100, consisting of just a few of the latest G-4 and He 219A nightfighters.

Below: During the closing months of the Second World War many Bf 110G-4s were hit by Allied fighter-bombers, or were destroyed by the Germans themselves because of a lack of aircraft fuel following the destruction of the main refineries in central Germany and elsewhere by Allied bombing. On this airfield lie the burnt-out remnants of several Bf 110Gs of NJG 3.

Top right: Despite the lack of AI radar installation, early Ju 88C heavy fighters such as this proved very successful in the early days of their operational careers. Besides those of NJGs 1 to 4 and 6, some Ju 88Cs were used by training units NJG 101 and 102. Several of the early aircraft had rebuilt bomber frames, which were fitted with a new nose section housing fixed armament.

Right: Its superior navigation equipment meant that the Ju 88C was used for dawn and night missions. This aircraft belonged to IV/NJG 5, which operated under the command of *Fliegerführer Atlantic*, and was used both to protect German Navy vessels and for missions against British long-range maritime aircraft over the English Channel and the Atlantic. Despite many operational successes the Ju 88 was only employed in this role in limited numbers, most such aircraft being used to strengthen night fighter units.

Left: The Ju 88C-4, which was in service from 1941 to late summer 1944, was powered by either two Jumo 211 B or two C piston engines. Its forward armament consisted of five MG 15s and one 20 mm MG FF aircraft cannon. Many C-types were called *Mischausführungen* because they were rebuilt bomber aircraft or modified heavy fighters.

Below: The Ju 88C was used for long-range night intruder (*Langstrecken Nachtjäger*) duties for a short period, and between 23 July 1940 and 13 October 1941 had been responsible for the loss of many Blenheims, Wellingtons and Hampdens, which were shot down as they approached their airfields on returning from missions over the Continent.

Right: A view inside a Ju 88C cockpit, possibly a C-4 night fighter. Beside the pilot's seat the heavy aircraft guns were situated side by side. The older MG FF guns fitted in nearly all C-series Ju 88s were replaced in G-1 and G-6 nightfighters by the more powerful MG 151/20.

Below: This crashed Ju 88C has been well camouflaged by its ground crew to provide protection against fighter-bomber attack. The aircraft was one of the *Mischtypen* or 'mixed types', a cross between the Ju 88C-4 and C-6 arrived at by using assorted spare parts. Many late *Luftwaffe* combat aircraft were pieced together from parts belonging to different series and were not listed in any official handbook.

Left: Ju 88C nightfighters and night intruders lacked the AI radar installations which would have enabled them to get more contacts with enemy aircraft. The engines installed later provided the improved performance necessary to fulfil their role against the four-engine Lancaster bombers which were appearing over the Reich in ever larger numbers.

Bottom left: This Ju 88C-6 was part of IV/NJG 1 and was captured by Allied forces early in 1945. It was one of the Ju 88 nightfighters which received an SN 2 antenna array. It represented the last development phase before the Ju 88G-1 became available, and was delivered to all important night fighter units throughout the Reich. In 1945 many of those Ju 88Cs still operational were used as heavy fighter-bombers to attack Allied columns which had crossed the Rhine.

Top right: Sitting on one of the BMW 801 radials of an NJG 100 Ju 88G-1 night fighter, this mechanic is clearly delighted to have received a letter from home.

Right: This Ju 88G-1, pictured before taking off for its home base after a nocturnal mission, was part of NJG 3. The Ju 88G-1 could achieve a maximum speed of more than 525 km/h at an altitude of some 6000 m. Standard armament consisted of four MG 151/20s, fitted in a large bulge situated under the main part of the fuselage.

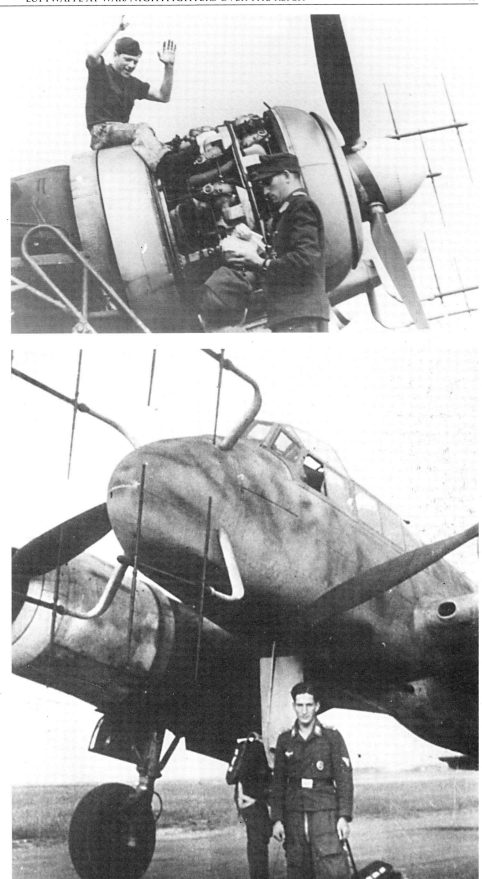

Above: Another Ju 88 G-1. Initially such aircraft were equipped with six 20 mm weapons, of which two could be installed in the forward fuselage. Because of the considerable weight of this heavy night fighter, it was subsequently decided to install only four of these guns. In addition there were plans to fit Kramer X-4 air-to-air missiles under both inner wings. However, the X-4 had not reached operational status in 1944, and its projected use was abandoned.

Below: This wreckage of a Ju 88G-1 (G9+MR, s/n 713521) was captured by American ground forces early in 1945. The aircraft had belonged to 7 *Staffel* of NJG 1 and was damaged by Allied night intruders.

Above: NJG 2 crews posing in front of one of their aircraft, a Ju 88G-1. By early 1945 as many as 2235 Allied bombers and long-range fighters had been shot down by German nightfighters. A further eighty-five Allied aircraft were missing in action and had been written off by the RAF.

Below: This Ju 88G-6 of 7/NJG 100 was fitted with one MG 131, with a flame-damper, in the rear defensive position. In addition there was an MG 151/20 *Schräge Musik* installation. The fixed forward-firing weapons positioned in the large lower bulge consisted of two MK 108s and two more MG 151/20s. A FuG 350 was situated on the rear cockpit fairing.

Above: An unidentified night fighter crew stand before their Ju 88G-1 (D5+BH), which belonged to 1 *Staffel* NJG 3. The fuselage is intensely camouflaged in three different shades of grey.

Below: Because of a lack of fuel this Ju 88G-6 is being pulled by four pairs of oxen, assisted by men belonging to the ground crew. All units were forced to save as much fuel as possible from late summer 1944 in order to make the most of their limited resources. This photograph was taken on one of the large airfields situated in central Germany.

Above: A captured Ju 88G-1, whose crew had landed on an RAF airfield in Britain in order to surrender. Only a few German pilots or crews flew abroad to hand over their aircraft to Allied authorities. Some additional aircraft landed on British airfields in consequence of pilot's error — 4R+UR, for example, a Ju 88G of 7/NJG 2.

Below: This severely damaged Ju 88G-1 night fighter of NJG 2 was demolished by retreating German forces in 1945. In the closing months of the war NJG 2 could only get enough fuel to carry out a handful of low level raids against ground targets.

Above: A captured Ju 88G-6, possibly photographed at Prague-Rusin, where a lot of former *Luftwaffe* aircraft and *Wehrmacht* vehicles were collected for scrapping. Many more Ju 88G-1s and G-6s were captured at Prague-Gbell and in the region round Saaz and near Pilzen. Early in May 1945 some Ju 88G nightfighters were damaged or destroyed by Czech insurgent forces.

Below: This Ju 88 G-6 fitted with flame-dampers was abandoned by German forces at Fassberg airfield. Often all usable spare parts of abandoned aircraft were first cannibalised, and if there was enough time complete engines were rebuilt and brought away, to be used in those Ju 88Gs withdrawn to central Germany in airworthy condition.

Above: D5+UK was part of 2 *Staffel* of NJG 3, and was destroyed by its own ground crew to prevent its capture by Allied forces. By the end of hostilities only a few Ju 88G-1s and G-6s, but no G-7s, remained in serviceable condition. These aircraft were used to evacuate *Luftwaffe* personnel from East Prussia to the northern part of Germany still held by the *Wehrmacht*.

Below: Between January and April 1945, successful sorties by British and American night intruders and fighter-bombers resulted in a dramatic reduction in the number of operational *Luftwaffe* nightfighters. Most of the surviving Ju 88s were dispersed in woods near their former airfields, or along one of the *Reichsautobahnen*.

Left: Some slightly damaged but unserviceable Ju 88G-6s were captured at Wunstorf airfield west of Hanover. This night fighter unit was engaged flying desperate low level attacks and a few night sorties early in 1945. What appears to have been the last German night fighter mission was an attack on advancing Allied tanks near the Mittelland Canal.

Bottom left: The Ta 154, Germany's 'wooden wonder', was just one of many ill-fated aircraft designs proposed for use by *Luftwaffe* night fighter units. This is the prototype, the Ta 154V1 (TE+FE), which took off on 1 July 1943 from Langenhagen near Hanover. During the first part of its evaluation it was flown without flame-dampers, in order to achieve a better performance than when fully equipped.

Above: The Ta 154 'Mosquito' was intended to be used as a fast night fighter, using either FuG 202 or Fu 220 (SN-2) AI radar with antennae fitted in the nose section. However, because of numerous problems related to finding the right wood glue, and with difficulties arising from the combination of metal and wooden parts, it did not become serviceable earlier than autumn 1944.

Below: This Ta 154A-4 (D5+HD, formerly KU+SU) belonged to a small batch of some five aircraft which saw operational use with NJG 3, based at Stade. After the loss of several Ta 154 nightfighters the remaining aircraft were transferred to the defence of northern Germany against British reconnaissance aircraft, especially the fast De Havilland Mosquito. However, there is no evidence that a Ta 154 ever shot down any RAF aircraft in that region.

Above: Besides using Ta 154A-4s as part of the *Reichsverteidigung* (Reich defence), one or two were handed over to *Ergänzungsjagdgeschwader* (Reserve Training Wing) EJG 2, which was based at Lechfeld near Landsberg/Lech in Bavaria. These were used to train young pilots selected to fly the Me 262 but lacking any experience in flying twin-engine fighters.

Below: The He 219 is believed to have been the most powerful two-seat night fighter of the Second World War. This is the first prototype (He 219V1, VG+LW, s/n 219001), tested by the Heinkel works from 6 November 1942. Because it was unarmed and did not carry full operational equipment it achieved a very high performance.

Above and below: This He 219V2 is GG+LW (s/n 219002, *Luftwaffe* operational designation G9+FB), which first took off on 10 January 1943. On 15 January it was flown by the well-known *Luftwaffe* ace, *Major* Werner Streib (sixty-six kills), who shot down five RAF bombers but crashed on landing as he returned from his successful sortie. Both pilot and wireless operator survived, but G9+FB was written off.

Top left: This photograph was taken during the final assembly of the pre-series He 219A-0. It shows a well-shaped cockpit section, in which two ejection seats were housed. Either side of the main fuselage were two compartments, each large enough to house a 20 mm cannon.

Bottom left: This He 219A-0, still without armament and yet to have its AI radar installed, has just left the final assembly line at Oranienburg near Berlin. The aircraft differed little from the later A-2 variant and was propelled by two DB 603 A piston engines. The large production line at Marienehe, near Rostock on the Baltic coast, was later bombed by Allied forces to prevent large-scale production of the new night fighter from getting under way.

Above: Most of the He 219A-0, A-1 and A-2 aircraft produced were transferred to NJG 1, where these powerful nightfighters were operated by three experienced officers, Streib, Hermann Förster and Ernst Wilhelm Modrow. Only a few He 219s were handed over to NJGr 10, an experimental night fighter unit, and *Nachtjagdstaffel* Finland, which was redesignated *Nachtjagdstaffel* Norway after Finland switched sides and obliged the Germans to withdraw their forces from its territory.

Below: This damaged He 219 belonged to I (*Einsatz*)/NJG 1, which was formed from surviving elements of the former three *Staffeln* of I/NJG 1 during the closing weeks of the war. The latter had operated more than forty aircraft in 1943–4, but the *Einsatzstaffel* consisted of only a handful of serviceable He 219s.

Left: This He 219A-0/R2 was previously equipped with a late version of FuG 220 radar, otherwise known as SN-2, since it bore under the cockpit the Roman number VI, identifying the radar variant installed in an individual night fighter. This aircraft was undergoing trials at Werneuchen, where electronic warfare equipment was prepared and tested.

Bottom left: Mock-up of a proposed new Ju 188. Many such mock-ups and a lot of different production designs were put together to rouse the interest of the RLM, but lack of capacity rendered it quite impossible to put large-scale production of new nightfighters in hand, even though the number of new parts that needed to be substituted was often relatively low.

Top right: This model of the projected Ju 388J-1 heavy night fighter shows a combination of older FuG 212 and new FuG 220 radar sets. The fixed armament was to comprise four 20 or 30 mm aircraft cannon, MK 103, MK 108 or the standard MG 151/20. A remote-controlled MG 131 Z was installed in the tail to protect the aircraft from attack in the rear.

Right: Because the interest of the relevant officers had been aroused, Junkers was allowed to construct this full-size mock-up of the Ju 388, with FuG 212 antennae. In addition to four fixed weapons in a bulge situated under the fuselage, it was intended that two *Bordkanonen* BK 3.7s would be installed forward.

Left: Another Ju 388 mock-up showed the combination of FuG 212 and FuG 220 also presented to the RLM. The full-scale model presented a new kind of installation, the antennae of the SN-2 radar. In spite of the great effort that went into the project, only a few Ju 388s were actually constructed in 1944–5 and none saw any operational service.

Below: The second Ju 388 prototype, designated V2 (s/n 500002), was handed over to *Erprobungskommando* 388, where many of the few Ju 388s produced were transferred for evaluation. This high-altitude night fighter was propelled by two supercharged BMW 801 TJ radials and was equipped with a pressurised cabin section housing a crew of three.

Left: A view into the rear part of a Ju 388 cockpit. Sitting on the right side, the wireless operator would handle the remote-controlled MG 131 Z, a development called FA (*Fernantrieb*, meaning remote-drive). Wireless sets were fixed on the rear wall.

Below: Because the remote-controlled FA gun system did not meet the *Luftwaffe*'s standards, a lot of extra work was called for to improve the aiming system and most of the remote functions. This full-scale model was constructed by Junkers workers to plan the installation of the FA 15 system.

Above: Two of the most effective design studies to be worked on after late 1944 were for the twin-engine Do 335 night fighter. The first was based on the Do 335B heavy destroyer, of which a few pre-series aircraft were built and later evaluated at Mengen and Friedrichshafen in 1945. The first experimental night fighter prototype (Do 335 V10) was completed in December 1944 and captured by Soviet ground forces at Oranienburg airbase. This view shows the cockpit interior of the series prototype.

Left: The second Do 335 night fighter prototype, designated M16, was manufactured at Ravensburg-Höll and later tested at Mengen; the cockpit shows that the fuselage was taken from a destroyer aircraft. It was propelled by two DB 603 engines, the forward installation belonging to the E-1, the rear to the LA, production series. This experimental aircraft was transferred to Bretigny near Paris after capture at the end of the war and was scrapped early in 1949.

Above: The Do 335 M16 first took off on 3 November 1948 and was tested by R. Receveau as an all-weather fighter until 1 June 1949. As a result of several accidents and the lack of spare parts evaluation was subsequently abandoned.

Below: This is Bindlach near Fürth, where the conversion of former Do 217A-1s and A-2s to nightfighter standard was carried out. The first fifty Do 335A-6s, all of them modified fighter-bombers, were intended to receive FuG 220 radar installations and did not differ from the Do 335 V10 in many regards. They were quite similar to two-seat training aircraft but flew with a FuG 218 radar installation.

Left: Only a few — possibly not more than six — He 111H-20 medium bomber transports were converted into training aircraft with an SN-2 installation. Three of them were used by front line units to train new wireless operators for night fighter formations. The large bulge under the middle of the H-20's fuselage is interesting, its purpose being unknown.

Opposite page, bottom: Unlike most of the twin-engine nightfighters, all the Bf 109G-6s of 2/NJG 300 remained in a dark camouflage. These aircraft were equipped with two additional MG 151/20 guns mounted beneath the wings. The aircraft depicted, photographed at Hangelar in August 1943, was flown by *Oberfeldwebel* Döring.

Below: From time to time experienced pilots of III/JG 300 were engaged in nocturnal missions over the Reich. This FW 190A-8 belonging to one of the *Sturmstaffeln*, together with twin-engine nightfighters of NJGs 3, 5 and 6 — all of them commanded by *Abschnittsführer Mittelrhein* — was used to attack heavy bombers in April 1944.

Only a handful of FW 190A-6s were flown with a FuG 217 'Neptun' radar installed. This aircraft, fitted with four antennae, was flown by *Oberfeldwebel* Migge, serving with 1 *Staffel* of NJGr 10 at Werneuchen. Besides this formation, many single-engine nightfighters were concentrated in the hands of NJG 11's pilots, one of whom was *Leutnant* Kurt Welter, who became the commanding officer of the Me 262-equipped 10/NJG 11.

Right: This 'Black 8', an FW 190A-6 with FuG 217 J, was part of I/JG 300 and took part in the costly night sorties undertaken during autumn and winter 1943. Because of the improved capabilities of the RAF's long-range nightfighters, and poor weather conditions, many of these fighters were lost in action or when returning to base. To increase their range, albeit insufficiently, one 300-litre drop tank was fixed under the fuselage

Right: In addition to the FW 190A-6, some BMW 801 D-powered A-5 and A-6 variants were also modified to become single-engine night-fighters. Their armament consisted of two MG 131s and four heavy MG 151s. Maximum speed was stated to be more than 640 km/h at an altitude of 8500 m, while maximum altitude attainable was 10,300 m. The unit badge consisted of a wild boar leaping through a star-spangled night sky.

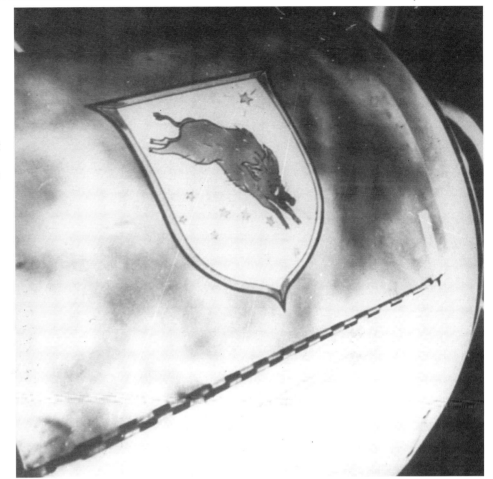

Above: In addition to the FW 190As, Bf 109G-6s and G-10s engaged in night fighter duties, a few FW 190D-9s such as this were handed over to JG 300 and JG 301 to improve combat performance. The lack of AI radar installations had resulted in most of them being used as day fighters, operating over the Reich and the Eastern Front.

Left: The first experimental Me 262A-1a (s/n 170056), fitted with FuG 218 radar. Tested by *Flugkapitän* Baur on 9 March 1945, it belonged to a limited series of rebuilt Me 262A-1a day fighters and prototypes used to undertake difficult night trials. During October and November 1944 *Oberst* Hajo Herrmann and *Major* Otto Behrens tried to apply the operational principles of single-engine nightfighter tactics to jet-propelled aircraft.

Above: Me 262 170056 — shortened to V056 — had never been used in action with AI radar equipment by the time the war ended. This aircraft was amongst the experimental Me 262s stationed at Lechfeld, which had become an important proving ground for all new Messerschmitt designs as well as being a *Luftwaffe* base.

Below: Me 262B-1s were very rare training aircraft, providing the initial step towards two-seat night fighter development. The evaluation of a two-seat jet-propelled night fighter aircraft was issued in summer 1944, based on the proposed two-seat trainer, the Me 262B-1a. One of the first design studies, called *Projektübergabe Me 262 Nachtjäger — Umbau Schulflugzeug Stufe I* (first stage of Me 262 night fighter rebuilt from a series jet fighter), was completed on 5 October 1944 and handed over to the RLM.

Above: The interior of the Me 262B-1a/U1 night fighter, of which the first was delivered to *Kommando* Welter in March 1945. This was the first jet fighter unit established specifically to fly night sorties. The B-1a/U1 achieved only a low level of success because of its many technical faults, most of *Oberleutnant* Welter's pilots preferring the Me 262A-1a without AI radar.

Below: All these aircraft belonged to the first and only Me 262-equipped night fighter unit, I/NJG 10. In eighteen missions with I/NJG 10 and 10/JG 300, Kurt Welter claimed twelve enemy aircraft shot down, followed by a further twenty-two after he was awarded the Knight's Cross. At the end of 1945 he became commanding officer of 10/NJG 11. His final score is unknown, but is thought to have been about fifty.

Above: 10/NJG 11 commenced action in February 1945 at Burg near Magdeburg. In March 1945 a small command operated from Oranienburg near Berlin, until both parts were reunited and withdrawn to Lübeck-Blankensee. Only a few weeks later most of Welter's surviving aircraft were handed over to the RAF at Schleswig-Jagel on 7 May 1945. The first of these was transferred to England on 18 May 1945.

Right: Despite many variant design suggestions during the summer of 1944, only two or three Ar 234B-2s were ever converted into nightfighters, called 'Nachtigall' (Nightingale). This first experimental Ar 234 night fighter (s/n 140145) was completed by the end of 1944 and was handed over to *Kommando* Bisping, established by order of *Luftwaffe* supreme command (or OKL, *Oberkommando der Luftwaffe*) on 12 December 1944.

Above: Aircraft like this Ar 234 were rebuilt into two-seat nightfighters, but when used over central Germany had little success against the more manoeuvrable De Havilland Mosquito bombers and nightfighters which operated there in significant numbers.

Above: This aircraft belonged to the Ar 234C series. Because reflections caused by the front glazing of their cockpits had been declared to be dangerous, production of all Ar 234 night fighter variants with the exception of the C-5/N was stopped early in 1945, and five new series called Ar 234 P-1 to P-5 were developed. However, the course of the war thereafter was such that none of them became available to enter action with any *Luftwaffe* formation.